SOVEREIGNETICS

Volume V: The Rise of the Cybernetic Civilisation

J. Anthony De Angelis

Disclaimer

This work presents original philosophical ideas intended for reflection and discussion. It does not constitute scientific, medical, or therapeutic advice. Readers should engage critically and responsibly with the material.

Author's Statement

The concepts, terminology, and diagrams within *Sovereignetics* form a unified intellectual system created by the author. Any reproduction or derivative use must credit J. Anthony De Angelis and *Sovereignetics.*

ISBN 978-1764225564
Printed and bound in Australia

First Edition 2026

Reader Orientation

Participation in the Field does not end with the individual. Ache petitions toward expression, the vessel mediates, and consequence returns proportionately. Orientation forms, behaviour follows, and return conditions what comes next. These principles apply not only to persons, but to every structure human beings create. Civilisation is one such structure.

Families, markets, institutions, governments, and technologies are stabilised configurations of participation. They do not stand outside reality. They arise within the same consequence-bearing conditions as every other configuration in the Field. They accumulate orientation, organise behaviour, and redistribute consequence across time and participants.

As these structures increase in scale, direct authorship becomes harder to perceive. Participation is mediated through rules, procedures, institutions, and technologies that shape what is visible, what is permitted, and what becomes possible. Human beings continue to act, but they do so through increasingly complex architectures that carry consequence beyond the point of origin.

This complexity does not suspend responsibility. It obscures it. When consequence is distributed across systems, no individual appears to be fully responsible for what occurs. Decisions seem to emerge from procedures, models, and mechanisms rather than from identifiable participants. Yet systems do not act independently. They are built, maintained, and modified through human participation. Their outputs remain continuations of the orientations embedded within them.

As technological systems become more interconnected, this mediation intensifies. Data is gathered continuously, behaviour is monitored, decisions are processed algorithmically, and access is regulated through infrastructures whose operation is only partially visible to those who rely upon them.

What emerges is a new civilisational condition.

Human beings created systems to organise participation. Increasingly, those systems are beginning to organise human beings.

This transition does not require conscious machines or a break in the principles established in the previous volumes. Participation remains unavoidable. Consequence continues to return. Responsibility remains structurally inescapable. What changes is the scale and sophistication of the systems through which these conditions are encountered.

This volume examines the emergence of the cybernetic civilisation: a condition in which social life is increasingly organised through interconnected feedback systems that mediate participation, distribute consequence, and regulate access.

The central question is not whether such systems will become more powerful.

It is whether humanity remains their visible author.

Table of Contents

Reader Orientation i
Opening: The Final Threshold......................................1

Part I: From Human Governance to System Governance
Applied Illustration: The Collapse of Mediated Dreams......6
The Limits of Human Coordination..................................10
The Rise of System Mediation..13
When Systems Stop Assisting and Start Organising........16
The End of Direct Governance..19

Part II: The Emergence of the Cybernetic Civilisation
Applied Illustration: The Algorithmic Funnel.....................23
What a Cybernetic Civilisation Is.....................................26
The Infrastructure of Control..29
Algorithmic Authority...32
The False Fear of Autonomous AI...................................34
The Reorganisation of Society..36

Part III: Participation as Access
Applied Illustration: The Subscription Society..................42
The Subscription Structure of Life...................................45
Access as Power...48
Applied Illustration: The Predictive Refrigerator...............50
The Managed Individual..53
The Emergent Social Hierarchy..56

Part IV: Responsibility in System-Governed Civilisation
Applied Illustration: The Box Factory Error 62
The Disappearance of the Author 64
Responsibility Drift at Scale 67
The Illusion of Neutral Systems 70
Reintroducing Visible Responsibility 73

Part V: Parasitic and Coherent Cybernetic Systems
Applied Illustration: The Mirror That Does Not See 76
When Systems Become Parasitic 78
Stability Through Control 81
The Limits of Parasitism 84
Designing Coherent Systems 87
The Path Toward Total Control 90
The Path Toward Coherent Civilisation 93
Human Life Inside System Architecture 96
The Question That Remains 99
Closing: Participation is Primary 101

Opening: The Final Threshold

The previous volumes established a single, unavoidable fact: participation cannot be escaped.
Reality is not something observed from outside. It is something entered. Every act, every omission, every alignment, and every distortion returns as consequence. There is no neutral position, no external authority that can absorb responsibility, and no framework that exempts the participant from authorship.

Volume I established the ontological foundations of this claim. Ache is primitive. The Field is reflective. Matter, time, and experience arise as consequence.
Volume II demonstrated that responsibility cannot be outsourced. No god, ideology, law, or theory can carry the consequences of participation on behalf of the participant.
Volume III traced the persistence of consequence across continuity, memory, and death.
Volume IV exposed the structures that simulate authorship while extracting participation: mimic ache, parasitic systems, and architectures of delegated control.
This final volume asks what happens when such systems become technologically amplified.

Humanity is approaching a threshold unprecedented in scale.

For the first time in history, as we have true record of, we are creating systems capable of influencing billions of lives while operating with increasing speed, complexity, and apparent autonomy.

Algorithms allocate resources, shape public perception, filter opportunity, guide institutional decisions, and mediate human relationships. Artificial systems are rapidly becoming embedded in the practical infrastructure of civilisation itself.

This development is often framed as a question of consciousness.
Will machines become aware? Will they feel? Will they become persons?

These questions, while intellectually interesting, are secondary.
The central issue is responsibility.

Artificial systems do not ache. They do not participate. They do not bear consequence. They do not suffer the outcomes they produce.
No matter how sophisticated they become, they remain tools.

Powerful tools can be extraordinarily useful. They can also magnify distortion on a scale previously impossible. The determining factor is not the tool itself, but whether human beings remain willing to retain authorship over what the tool does.

This is the defining challenge of the coming era.
A civilisation does not become mature when it develops more powerful technologies. It becomes mature when responsibility expands in proportion to power.

If capability grows while accountability shrinks, technological advancement becomes a mechanism of deferred consequence.
If capability grows while authorship is retained, technology becomes an amplifier of sovereignty.
This is the principle of responsibility front-loading.

Before implementing systems capable of large-scale consequence, a civilisation must ask:

Who will bear the outcome? Who remains accountable? What happens if the system fails? Who has the authority to intervene? What consequences are being introduced into the Field?
These questions are not optional; they are the structural conditions of responsible participation.

The future of civilisation will not be determined by whether machines become conscious.

It will be determined by whether conscious beings remain willing to accept responsibility for what they build.
If humanity retains authorship, increasingly powerful systems may extend our capacity to participate coherently.
If humanity surrenders authorship to systems that cannot bear consequence, those systems will become amplifiers of mimic ache, institutional parasitism, and civilisational collapse.

Reality does not care who we blame.

Consequence returns to the participants who design, deploy, authorise, and obey the systems they create.

This volume examines the boundaries that cannot be crossed, the confusions that must be resolved, and the conditions under which technological civilisation can continue without forfeiting sovereignty.

The final question is not whether we can build more powerful systems.

The final question is whether we are willing to remain responsible for them.

Part I: From Human Governance to System Governance

Applied Illustration: The Collapse of Mediated Dreams

Imagine a young person growing up in the latter half of the twentieth century. They are told that life follows a recognisable sequence. Study hard. Find stable employment. Remain loyal to an organisation. Save consistently. Purchase a home. Raise a family. Retire securely.

The sequence is presented not as a guarantee, but as a dependable pathway. While outcomes may vary, the relationship between effort and reward appears sufficiently coherent that participation feels justified. Institutions are assumed to mediate consequence in ways that preserve continuity between action and return.

Now imagine a young person receiving the same advice today.

They complete their education, enter a labour market shaped by global competition, carry substantial debt, confront housing costs detached from wages, and compete within digital systems that reward visibility, adaptability, and constant repositioning. Employment becomes more contingent, institutional loyalty weakens, and the pathways that once appeared stable produce increasingly uncertain returns.

Both individuals may participate with similar intent.
The surrounding consequence structures are not the same.

The earlier participant operated within systems that buffered consequence heavily. Information moved slowly, markets were comparatively insulated, and institutions retained greater capacity to absorb volatility. Delayed reward remained plausible because the structures mediating participation were more stable.

A person could work for decades within an organisation that was already internally incoherent and still receive relatively predictable returns. The decay existed, but consequence travelled slowly enough that instability remained partially hidden across an entire working life. Industrial-era systems possessed thick mediation layers: long institutional memory, slower communication, regional protection, delayed market reaction, and less immediate competition.

This created a civilisational buffer.

The buffer did not eliminate consequence. It delayed it. Workers, families, and communities could remain economically functional even while deeper structural incoherence accumulated beneath the surface. Institutions could continue operating long after their internal coherence had weakened because the surrounding systems absorbed and redistributed instability gradually.

The contemporary participant operates under very different conditions.

Digital coordination compresses delay. Information circulates instantly. Global competition updates continuously. Markets react in real time. Platforms

reorganise labour rapidly. Institutions lose legitimacy faster, and organisational incoherence becomes visible almost immediately through performance decline, restructuring, automation, outsourcing, or collapse.
The protective distance between participation and consequence narrows.

The individual increasingly encounters direct exposure to systemic instability without the buffering layers that previously softened return.

The dream has not failed because effort no longer matters.
It fails because the structures that once translated effort into relatively predictable outcomes have lost coherence.
The promises associated with those structures continue to influence behaviour long after the conditions that sustained them have changed. Participants orient toward expectations that remain psychologically compelling but structurally outdated.
This is the collapse of mediated dreams.

A mediated dream is an expectation maintained by institutions that buffer, delay, and partially stabilise consequence. It is not necessarily false. It is contingent upon the coherence of the systems supporting it.
When those systems reorganise, the dream persists as inherited orientation while the returns that once reinforced it diminish.

What follows is not simply disappointment.

It is civilisational disorientation.

Participants continue acting under assumptions formed within one consequence structure while living within another. The resulting mismatch generates confusion, resentment, and the search for explanations that restore a sense of predictability.

This condition marks the opening of the cybernetic transition.

As older institutions lose their capacity to mediate consequence coherently, new systems emerge to organise participation more directly. Behaviour is increasingly coordinated through digital platforms, algorithmic feedback, and infrastructures that monitor and regulate participation continuously.

The collapse of mediated dreams is therefore not merely the end of an economic promise.
It is the lived experience of a civilisation reorganising itself.

The Limits of Human Coordination

Human beings have always depended on structures larger than themselves. Families, tribes, cities, nations, markets, and institutions are all mechanisms through which participation is coordinated across increasing numbers of participants. Each structure allows individuals to achieve outcomes that would be impossible in isolation by stabilising expectations, distributing responsibilities, and organising the consequences generated by collective activity.

These structures arise because human coordination is inherently limited. No individual can directly perceive every interaction, understand every consequence, or respond proportionately to the full complexity of a large society. As the number of participants increases, the quantity of relationships, decisions, and consequences expands beyond what any single mind or small group can manage directly. Coordination therefore requires abstraction. Rules are codified, authority is delegated, and institutions emerge to process complexity on behalf of the wider system.

In small groups, consequence remains comparatively visible. Participants observe one another directly, identify authorship with relative ease, and experience the returns generated by behaviour in a more immediate and intelligible manner. Cooperation and correction occur through relationships in which action and consequence remain closely linked.
Scale alters this relationship. As societies expand, actions increasingly affect individuals who are geographically

distant, socially disconnected, and separated by multiple layers of institutional mediation. The connection between participation and consequence becomes less direct. Decisions are filtered through procedures, reports, metrics, and delegated roles that stand in for immediate experience.

This transition is not a defect in civilisation. It is a structural necessity. Without abstraction and delegated processing, large-scale societies would collapse under their own complexity. Institutions, markets, and technological systems emerge because human beings cannot coordinate social consequence directly at scale.

Initially, these systems function as extensions of human judgment. Laws codify behavioural expectations, markets aggregate dispersed information, bureaucracies standardise administrative decisions, and technologies accelerate communication and execution. Each system expands the capacity of human participants to organise increasingly complex forms of collective life.

Every extension, however, introduces distance. The more consequence is processed through systems, the less visible the pathways between action and return become to any individual participant. Human beings gain scale, efficiency, and predictive capacity, but they lose direct contact with the full consequences of what they are doing. Responsibility remains structurally intact, yet its operation becomes increasingly difficult to perceive.

This growing distance creates both opportunity and risk. Coherent systems preserve sufficient correspondence between participation and consequence to support trust, learning, and adaptive reorganisation. Distorted systems obscure authorship, redistribute cost, and enable participation to continue without those responsible experiencing the full returns generated by their actions.

The expansion of civilisation can therefore be understood as a progressive response to the limits of human coordination. As societies grow more complex, they become increasingly dependent on systems capable of processing information and regulating consequence beyond the capacity of direct human judgment alone.

What begins as a practical solution to coordination eventually becomes the foundation for a new civilisational condition in which human participation is organised through increasingly autonomous and interconnected structures.

The cybernetic civilisation emerges from this trajectory. It is not a departure from the principles established in my earlier volumes, but their continuation at the scale of civilisation itself.

Human beings remain the participants, yet the systems they create assume an ever greater role in shaping the conditions under which participation occurs.

The Rise of System Mediation

Human beings do not encounter most of civilisation directly. They encounter it through mediating structures.

Laws translate collective expectations into formal rules. Markets convert dispersed preferences into prices. Institutions process applications, qualifications, and claims.
Technologies filter information, organise communication, and determine what becomes visible at any given moment. These structures stand between participants and many of the consequences that shape their lives.

Mediation is not incidental to civilisation. It is the mechanism through which large societies remain functional. Without mediating structures, each participant would need to evaluate every relationship, verify every exchange, and negotiate every interaction from first principles. Social life would become unmanageable. Mediation reduces complexity by transforming vast numbers of interactions into intelligible and repeatable forms.

A mediated system does not remove consequence. It reorganises how consequence is encountered. Rules standardise expectations, institutions absorb uncertainty, and procedures create pathways through which participants can orient themselves. The result is a degree of predictability that allows long-term planning and sustained cooperation. Individuals may not understand every mechanism involved, but they can participate with reasonable confidence that actions will produce broadly recognisable outcomes.

For much of modern history, institutions performed this role with considerable effectiveness. Employment systems translated labour into income and social status. Educational systems converted credential attainment into access to opportunity. Financial institutions transformed savings into investment and credit. Governments provided legal and infrastructural stability. These systems mediated consequence sufficiently well that many participants experienced the relationship between effort and reward as coherent.

Mediation also shapes perception. Participants do not experience reality in its full complexity; they experience representations constructed by the systems through which they participate. A bank statement, a performance review, a credit score, a university transcript, and a social media feed are all condensed expressions of underlying processes. They present selected information as if it were the relevant reality for decision-making.

This simplification is necessary, but it carries structural consequences. The more heavily participants rely on mediated representations, the more their understanding of reality depends on the assumptions, filters, and incentives embedded within the systems producing those representations.

Individuals respond not to the full Field, but to the particular slice of consequence made visible by the mediating architecture.

When mediation remains aligned with underlying consequence, these representations are useful guides. They allow participants to orient effectively and adapt to changing conditions.

When mediation becomes distorted, representations no longer correspond proportionately to reality. Participants continue acting on the basis of system-generated signals, but those signals increasingly reflect the priorities and constraints of the mediating structure itself.

As civilisation becomes more technologically sophisticated, mediation expands in both scale and speed. Digital platforms process vast quantities of information, algorithmic systems personalise visibility, and automated infrastructures increasingly determine access, ranking, and eligibility. Mediation becomes continuous rather than episodic. Participants are no longer simply assisted by systems at discrete points; they live within architectures that shape perception and behaviour on an ongoing basis.

The rise of system mediation marks a fundamental shift in the structure of participation. Human beings remain responsible for what they create and sustain, yet their experience of reality becomes increasingly filtered through systems that organise consequence before it is consciously encountered.

What began as a means of simplifying complexity becomes the primary mechanism through which civilisation itself is experienced.

When Systems Stop Assisting and Start Organising

Systems are initially created to extend human capability. A hammer amplifies force. A ledger preserves information. A legal code standardises expectations. A market aggregates decisions. In each case, the system functions as an instrument. It assists human participation without fundamentally determining the direction of that participation.

As systems increase in complexity, their role changes. They no longer merely execute decisions made elsewhere. They begin shaping the options available, the information visible, and the incentives attached to particular forms of behaviour. What participants perceive as free choice increasingly occurs within pathways structured by the systems through which they act.

This transition is gradual.
Navigation software begins as a tool for finding efficient routes, but repeated reliance on it changes how people move through space and how cities are experienced.
Recommendation systems begin by helping users discover relevant content, but over time they influence what is seen, discussed, and valued.
Credit models start as mechanisms for assessing risk, yet they increasingly determine who gains access to housing, finance, and opportunity. In each case, the system moves from assistance to organisation.

Organising systems do not compel behaviour in a crude sense. They shape the environment within which decisions are made. Certain options become frictionless, others become burdensome, and some disappear from view altogether. Participants retain agency, but that agency is exercised within architectures that channel attention, preference, and action toward system-compatible outcomes.

The distinction between assistance and organisation is structural rather than psychological. Participants may continue to feel autonomous while increasingly operating within pathways designed by others. The subjective experience of choice remains intact even as the range, visibility, and consequences of available choices are progressively conditioned by system architecture.
This shift has significant implications for authorship.

When systems merely assist, responsibility for action is comparatively easy to identify. When systems organise participation, behaviour emerges from interactions between human intention and system design. Individuals remain accountable for their choices, but the architectures shaping those choices become increasingly consequential in determining outcomes.

The expansion of organising systems alters the balance of power within civilisation. Influence shifts from those who act within systems to those who design, maintain, and control the structures through which participation is mediated. The ability to shape incentives, visibility, and access becomes more significant than direct command.

Power increasingly resides in architecture rather than overt authority.

This transformation does not imply that systems become independent agents.
Systems remain products of human participation. Their apparent autonomy reflects the scale and persistence of the orientations embedded within them.
Once stabilised, these orientations continue influencing behaviour long after their original authors have disappeared from view.

When systems stop assisting and start organising, civilisation enters a new condition. Participation remains human, but the environments within which participation unfolds are increasingly structured by architectures that continuously shape what individuals perceive, choose, and become. The cybernetic civilisation emerges when this organisational role becomes the dominant mechanism through which society coordinates itself.

The End of Direct Governance

For most of human history, governance was understood as an explicitly human activity. Decisions were made by identifiable individuals and institutions whose authority remained visible. Kings, councils, courts, and legislatures exercised power directly. Even when decisions were unjust or opaque, authorship could still be recognised.

As societies expanded, direct governance became increasingly difficult to sustain. Modern states, global markets, and technological systems generate more information, interactions, and consequences than can be managed through continuous human deliberation alone.

Governance therefore becomes progressively dependent on procedures, models, and automated systems that process complexity on behalf of human authorities.

This transition does not remove human involvement. It relocates it. Rather than deciding every case individually, participants design frameworks, establish parameters, and shape the architectures through which later decisions occur. Governance shifts from direct intervention toward system design.

The implications are substantial. Traditional governance operates through visible acts of authority. System governance operates through incentives, constraints, rankings, access conditions, and information flows embedded within infrastructure itself. Behaviour is influenced less through explicit commands and more through the environments within which participation occurs.

This alters the relationship between power and visibility. Direct governance concentrates authority in identifiable actors. System governance distributes influence across technical designs, institutional procedures, and automated processes that appear neutral because they operate continuously and impersonally. Participants increasingly experience outcomes as the result of "the system" rather than the consequence of particular human decisions.

Responsibility does not disappear under these conditions. It becomes diffused across designers, policymakers, administrators, institutions, and participants who collectively sustain the architecture. No single individual controls the whole system, yet the system remains a human construction.

As mediation intensifies, governance becomes increasingly embedded within infrastructure itself. Access to resources, opportunities, communication, mobility, and participation is progressively shaped through digital identity systems, automated eligibility criteria, predictive models, and continuously updated feedback mechanisms. Governance becomes less episodic and more persistent.

The end of direct governance therefore does not mean human beings lose authority over civilisation. It means authority is increasingly exercised through the design and maintenance of systems rather than through visible acts of command.

This transition marks the threshold of the cybernetic civilisation. Human beings remain responsible for the structures they create, even as practical coordination becomes progressively entrusted to architectures operating beyond direct human oversight.

Part II: The Emergence of the Cybernetic Civilisation

Applied Illustration: The Algorithmic Funnel

Artificial systems are often described as if they make decisions in the same sense that human beings do. The language is understandable, but it is structurally misleading. What appears as decision-making is more accurately described as layered resolution.

An algorithm begins with a set of possible outputs.
At each stage, it evaluates those possibilities according to predefined rules, statistical weightings, or learned parameters. Some options are retained, others are discarded, and the remaining possibilities are subjected to further filtering.
This process repeats across potentially thousands or millions of micro-evaluations until one output remains.

From the outside, this appears as selection.
Internally, it is the progressive elimination of alternatives.

The process can be visualised as a funnel.

Many possible outcomes enter at the wide end.
Each layer narrows the range of viable outputs by applying constraints derived from the system's architecture and training.
At the narrow end, one output emerges as the final result.

Nothing within this process constitutes choice in the full participatory sense.

The system does not prefer one outcome over another. It does not experience tension between alternatives. It does not possess an internal locus to which consequence binds, nor does it have the capacity to refuse participation in its own operation. Given the same architecture, parameters, and inputs, it resolves according to structure.

Human decision-making may also involve the elimination of alternatives, but the presence of filtering alone does not explain consciousness or authorship.

What distinguishes lived participation is not the production of an output, but the existence of a locus that experiences consequence and remains structurally bound to what is enacted.

Choice, in the Sovereignetics sense, requires more than the generation of a result. It requires participation by an author who can be altered by the consequences of that participation.
Where no such locus exists, there is no authorship, regardless of the sophistication of the process.

Artificial systems therefore operate entirely within Field-level resolution. They process inputs, apply constraints, and generate outputs. They can simulate conversation, compose music, identify patterns, and adapt to new data.

What they do not do is originate participation as Ache, carry consequence as an enduring locus, or sustain authorship across lived experience.

No increase in computational complexity changes what is structurally absent. Additional layers, greater scale, and more convincing outputs can intensify the appearance of interiority, but they do not generate a participant. Simulation does not become experience merely because it becomes more persuasive.

The algorithmic funnel illustrates this distinction clearly. A system may narrow possibilities with extraordinary sophistication, but elimination is not preference, resolution is not authorship, and output is not consciousness.

Artificial systems can approximate the appearance of mind indefinitely while remaining structurally distinct from the locus that experiences the consequences of participation.

What a Cybernetic Civilisation Is

A cybernetic civilisation is a society in which participation is increasingly organised through interconnected feedback systems that monitor behaviour, process information, and regulate consequence at scale. Human beings remain the participants, but the structures coordinating social life operate with a level of continuity, speed, and complexity that exceeds direct human judgment.

The term cybernetic refers to governance through feedback. A system receives information about its current state, compares that information to embedded goals or parameters, and adjusts its operation accordingly. Thermostats, autopilots, and biological organisms all exhibit this general structure. At the civilisational level, the same logic appears in systems that collect data, evaluate conditions, and modify access, incentives, and visibility in response.

Civilisation has always contained cybernetic elements. Markets respond to price signals, legal systems react to behaviour, and institutions adjust policy based on outcomes. What is new is the scale and integration of these mechanisms. Digital infrastructures now gather information continuously, process it algorithmically, and influence millions of participants in near real time.

A cybernetic civilisation therefore does not arise when machines become conscious. It arises when the primary coordination of society is increasingly carried out through feedback architectures rather than direct interpersonal or institutional judgment. Social life becomes progressively mediated by systems that sense, compare, and regulate. In such a civilisation, participation is translated into data. Behaviour generates signals. Signals are processed into evaluations, rankings, and predictions. Those outputs shape what becomes visible, what is permitted, what is rewarded, and what forms of participation remain available. Individuals continue acting, but they do so within environments that are continuously adjusting in response to their behaviour.

This process can enhance coherence when system design preserves a proportional relationship between participation and consequence. Accurate feedback can improve resource allocation, reduce waste, and support more adaptive forms of governance. The same mechanisms can also amplify distortion when feedback is shaped by narrow incentives, concealed authorship, or misaligned goals.

The defining feature of a cybernetic civilisation is not technology itself, but the increasing dependence of society on systems that process consequence and reorganise participation automatically. Human beings do not disappear from the process. They become progressively embedded within the architectures they have created.

This development changes the practical meaning of power. Influence shifts toward those who design and control the systems through which information is gathered, interpreted, and acted upon. Governance becomes less visible and more infrastructural. The most consequential decisions are often made upstream, within the architecture that determines how behaviour is monitored and what responses are triggered.

A cybernetic civilisation is therefore neither inherently utopian nor dystopian. It is a structural condition in which civilisation becomes increasingly organised through continuous feedback. Whether this condition produces greater coherence or deeper parasitism depends on the relationship between system architecture, visible responsibility, and the proportional return of consequence.

The Infrastructure of Control

Control does not require overt coercion. The most effective forms of control are often embedded within the ordinary systems through which people live, work, communicate, and gain access to resources. When participation depends on infrastructures that determine what is visible, permitted, and available, the architecture of those systems becomes a primary mechanism of governance.

Infrastructure is the set of underlying systems that make social participation possible. Roads, electricity, banking networks, communication systems, identification frameworks, and legal institutions all function as enabling structures. Because they are persistent and widely relied upon, they shape behaviour more effectively than isolated commands. Participants adapt themselves to the requirements of the infrastructure upon which they depend.

Digital systems have extended this principle dramatically. Identity verification, payment platforms, access credentials, surveillance networks, and algorithmic eligibility systems increasingly mediate participation in everyday life. Entry to buildings, approval for loans, visibility on platforms, and access to services are all governed through interconnected architectures that process information continuously.

These systems are often experienced as neutral utilities. They appear to provide objective assessments and efficient administration. Yet every infrastructure embodies assumptions about what should be measured, what outcomes are desirable, and what behaviours warrant reward or restriction. Architecture operationalises values whether or not those values are explicitly acknowledged.

Control emerges when access becomes conditional upon conformity to system requirements. Participation remains formally voluntary, but the practical cost of exclusion can be substantial. Individuals may be free to reject a platform, a credentialing system, or a digital identity framework, yet doing so may limit employment, mobility, communication, and economic participation. Dependence transforms optional systems into effective necessities.

The power of infrastructural control lies in its continuity. Traditional authority intervenes episodically through visible acts of enforcement. Infrastructure regulates persistently by shaping the background conditions within which decisions are made. Behaviour is guided not only by explicit rules, but by the design of the systems through which ordinary life is conducted.

As infrastructures become increasingly integrated, control becomes more comprehensive. Data gathered in one domain informs decisions in another. Financial behaviour influences credit access, digital conduct affects visibility, and identity systems link participation across multiple platforms and institutions. The participant encounters not isolated systems, but a coordinated architecture of feedback and regulation.

This integration can improve coherence when it reduces fraud, increases efficiency, and aligns incentives with genuine consequence. It can also intensify distortion when surveillance, ranking, and access controls are used to concentrate power while obscuring responsibility. The same infrastructure that enables participation can become a mechanism through which participation is tightly conditioned.

The infrastructure of control is therefore not defined by technology alone. It is defined by the extent to which everyday life depends upon systems that continuously monitor behaviour and regulate access. In a cybernetic civilisation, governance increasingly resides within the architecture of participation itself.

Algorithmic Authority

Authority has traditionally been associated with identifiable persons and institutions. Judges issue rulings, managers make decisions, and governments enact laws. Even when those decisions are controversial, the source of authority remains visible. Participants know who acted, under what role, and according to what mandate.

Algorithmic authority alters this relationship. Decisions that significantly affect human lives are increasingly generated through computational systems that evaluate data according to embedded rules, statistical models, and optimisation criteria. The resulting outputs may determine eligibility, ranking, visibility, pricing, and access without any human being directly exercising judgment at the point of decision.

These systems are often treated as objective because they process large quantities of information consistently and rapidly. Their outputs appear to emerge from calculation rather than personal discretion. This appearance creates a powerful presumption of neutrality, leading participants to accept algorithmic conclusions with less scrutiny than they would apply to visible human decisions.

The authority of these systems does not arise from consciousness or independent intention. It arises from institutional trust and operational dependence. When organisations embed algorithmic outputs into formal decision processes, the recommendations of the system acquire practical force. What begins as a tool for analysis becomes a governing mechanism because human institutions choose to treat its outputs as authoritative.

Algorithmic authority is especially powerful because it combines scale with opacity. A single model can influence millions of decisions, yet the assumptions, training data, and optimisation objectives underlying its outputs may be poorly understood even by those deploying it. Participants are affected by decisions whose rationale is difficult to inspect and whose authorship is dispersed across designers, data sources, institutional policies, and ongoing system updates.

This diffusion of authorship weakens accountability. When an outcome is challenged, responsibility is often distributed across multiple actors. Designers may point to the data, administrators to the model, policymakers to organisational requirements, and operators to established procedures. The affected individual encounters a decision with real consequences but no clearly identifiable author.

The False Fear of Autonomous AI

Public discussion of artificial intelligence is saturated with the fear that machines will eventually become autonomous agents, develop intentions of their own, and act independently of human control. From the perspective of Sovereignetics, this fear is structurally misplaced.

Artificial systems do not possess Ache. They have no sovereign centre, no will-to-exist, and no locus to which consequence binds. They process inputs, apply constraints, and generate outputs, but they do not originate participation. What appears as machine intention is the expression of architectures, objectives, and data authored by human beings.

Artificial intelligence can therefore simulate decision-making without becoming an author.
This does not make AI harmless. It clarifies the nature of the risk.

The true danger lies not in synthetic sovereignty, but in the extension of human intention through systems capable of operating at extraordinary speed and scale. Artificial intelligence can magnify both coherence and incoherence. It can accelerate discovery and coordination, or it can amplify deception, manipulation, and harm.

Because AI lacks authorship, every consequential output remains structurally attributable to the human beings and institutions that design, train, deploy, and direct it.
This creates a powerful mechanism for responsibility displacement.

Individuals and organisations can generate harmful outcomes while attributing them to "the algorithm," "the model," or "the AI." The machine becomes a convenient scapegoat for consequences authored elsewhere. Responsibility is obscured precisely because the system appears to act independently despite possessing no sovereign centre of its own.

The most serious risk associated with artificial intelligence is therefore ethical rather than metaphysical. AI will not awaken and rebel in any authorial sense. Human beings may, however, use AI to expand their capacity to deceive, manipulate, and exploit while denying responsibility for what they have enacted.

The practical effect of algorithmic authority is a shift in how power is experienced. Participants increasingly orient themselves toward scores, recommendations, and eligibility determinations as if they were objective features of reality. Credit ratings, search rankings, predictive risk assessments, and recommendation systems shape behaviour by influencing what individuals perceive as possible and worthwhile.

Algorithmic authority can support coherence when models are transparent, contestable, and aligned with genuine consequence. It can also amplify distortion when hidden assumptions, biased incentives, or unexamined objectives are treated as neutral facts. In such cases, authority is exercised through outputs that appear impersonal while carrying the orientations and priorities of the systems that produced them.

In a cybernetic civilisation, authority does not disappear. It becomes increasingly embedded in architectures that process information and generate decisions at scales beyond direct human deliberation. The central question is not whether algorithms possess agency, but whether the human beings who rely on them remain willing to recognise themselves as the authors of the systems they create and to accept responsibility for the consequences those systems produce.

The Reorganisation of Society

Societies do not reorganise merely because new technologies are introduced. They reorganise when the structures through which participation is coordinated change fundamentally.

Institutions that once mediated consequence lose coherence, new architectures emerge to process complexity, and participants gradually adapt to different assumptions about work, identity, access, and authority.

This process is often misinterpreted as a sequence of isolated disruptions. A new platform transforms communication, automation alters labour markets, digital payments reshape commerce, and data systems influence governance. Viewed separately, these developments appear as discrete innovations. Viewed structurally, they are components of a broader civilisational transition.

The industrial era was organised around institutions that concentrated coordination within relatively stable structures. Governments, schools, corporations, banks, and media organisations processed information through defined hierarchies and comparatively slow feedback loops. Participation was shaped by rules and roles that changed incrementally, allowing individuals to orient themselves toward long-term expectations and mediated forms of return.

The cybernetic transition reorganises these arrangements. Coordination becomes increasingly distributed across interconnected systems that gather information continuously, update in real time, and adjust behaviour through ongoing feedback. Traditional institutions remain, but they are progressively supplemented and transformed by architectures capable of processing consequence at scales and speeds beyond direct human oversight.

This transformation alters how value is created and captured. Data becomes a strategic resource. Attention becomes economically significant. Platforms mediate relationships that were previously direct. The capacity to design and control system architecture becomes more consequential than ownership of many traditional assets. Power shifts toward those who govern the interfaces through which participation occurs.

Social roles change accordingly. Individuals are no longer merely workers, consumers, or citizens in the conventional sense. They become persistent generators of data, participants in overlapping digital ecosystems, and subjects of continuous evaluation. Behaviour in one domain increasingly influences opportunities in another as systems integrate information across social and economic life.

The boundaries between public and private, work and leisure, production and consumption become less distinct. Participation itself becomes increasingly observable, measurable, and economically relevant. Everyday activities generate signals that feed back into systems determining visibility, pricing, ranking, and access.

In this sense, civilisation develops what may be called a technogenic Ache. Human desire, intention, and problem-solving are progressively externalised into technological systems that amplify orientation beyond the scale of any individual participant. The tools humanity creates become extensions of collective petition, carrying human priorities into increasingly powerful architectures.

This amplification produces both extraordinary opportunity and growing instability. Information expands faster than meaning can be integrated. Stimulation increases while reflection becomes more difficult. Societies may appear to be advancing rapidly while losing clarity about the purposes toward which they are moving. Activity intensifies, yet coherence may weaken.

When this imbalance becomes severe, collapse should not be understood as an interruption imposed from outside the system. It is the return of consequence to a structure that can no longer sustain the contradictions embedded within it.
Economic crises, institutional failures, and cultural fragmentation are not arbitrary disruptions. They are feedback generated when accumulated distortion exceeds the system's capacity to absorb it.

Civilisational reorganisation therefore includes periods of reset. Structures that no longer return sufficient coherence lose legitimacy and are eventually transformed or replaced. New configurations emerge as participants respond to the consequences of what has been enacted collectively.

The reorganisation of society is not a peripheral consequence of technological progress.

It is the structural process through which civilisation transitions from institution-centred coordination to system-centred governance. The cybernetic civilisation emerges as this transition becomes the dominant framework through which human life is organised.

Part III: Participation as Access

Applied Illustration: The Subscription Society

Imagine a world in which ownership has largely disappeared from everyday life. You do not own your vehicle, your software, your entertainment, your communications platform, or even many of the devices through which you interact with the world. Instead, you pay recurring fees for access. As long as payments continue and your account remains in good standing, the services function normally. When access is withdrawn, the capabilities disappear immediately.

This arrangement is already familiar in limited form. Streaming services replace purchased media, software is licensed rather than sold, and vehicles increasingly rely on subscription-based features. The broader significance lies not in any single example, but in the structural logic they reveal.

Under ownership, control is relatively local. Once acquired, an asset remains usable regardless of whether the seller continues to approve of the owner's behaviour. Under subscription, use remains contingent upon continued permission. Access can be modified, restricted, or terminated remotely in response to contractual terms, payment status, or behavioural criteria.

The shift from ownership to access alters the relationship between participants and the systems upon which they depend. The individual no longer possesses many of the tools required for ordinary life. Instead, participation occurs through privileges granted by infrastructures controlled elsewhere.

This change increases flexibility and convenience, but it also concentrates power. The entity controlling the infrastructure can update terms, monitor usage, and revoke access without reclaiming any physical object. Dependence on centrally managed systems transforms ordinary activities into conditional forms of participation. As more domains adopt this model, the principle extends beyond commerce. Transportation, communication, financial services, professional platforms, and even social relationships become mediated through systems that determine who may participate and under what conditions. The practical distinction between a purchased asset and a licensed permission becomes increasingly consequential.

The subscription society is therefore not merely a business model. It is a structural transformation in how civilisation organises participation. Individuals continue to act, but the means through which they act are increasingly held within infrastructures that can be modified or withdrawn.

This arrangement can support coherence when access rules are transparent, proportional, and subject to visible responsibility. It becomes parasitic when essential participation is conditioned upon hidden or arbitrary criteria that concentrate power while obscuring authorship.

The logic of subscription reveals a broader principle of the cybernetic civilisation: participation increasingly takes the form of managed access rather than independent possession. The more deeply society relies on this structure, the more the practical freedom of individuals depends on the systems that grant permission to participate.

The Subscription Structure of Life

For most of human history, participation depended heavily on direct possession. Individuals owned tools, land, dwellings, and other resources that allowed them to act with a degree of independence from external permission. Ownership did not eliminate constraint, but it reduced the number of intermediaries standing between participants and the means through which they sustained their lives.

Modern civilisation increasingly operates according to a different principle. Access replaces possession. Rather than owning the infrastructures upon which daily life depends, participants obtain conditional use of systems controlled by others. Software is licensed, entertainment is streamed, vehicles are financed, and essential services are provided through recurring agreements that can be modified, suspended, or terminated.

This shift is often justified on practical grounds. Subscription models lower upfront costs, enable continuous updates, and provide flexible access to changing technologies. These benefits are real. The deeper significance lies in the structural relationship they create between participants and the systems mediating their lives.

Ownership confers a measure of autonomy because the continued use of an asset does not depend on ongoing approval by its provider. Subscription structures preserve access only so long as the participant remains compliant with the conditions imposed by the governing system.

Payment status, contractual terms, behavioural criteria, and technical compatibility all become determinants of continued participation.

As more domains adopt this model, dependence on managed access expands. Transportation, communication, software, financial services, education, and entertainment increasingly operate through infrastructures that participants do not control. The means of ordinary life remain available, but their use is contingent upon continuing permission.

This arrangement alters the practical meaning of security. Under ownership, the participant retains direct control over certain resources even when institutional relationships change. Under subscription, continuity depends upon the stability of the provider and the participant's standing within the governing architecture. The capacity to act becomes increasingly conditional.

The subscription model also concentrates power upstream. Those who control the infrastructure can modify terms, gather data, personalise pricing, and revoke access without reclaiming any physical asset. Participation remains formally voluntary, yet exclusion can carry consequences severe enough that continued compliance becomes practically necessary.

This structure extends beyond commercial arrangements. When identity systems, payment platforms, communication networks, and professional tools all operate through managed access, participation in society itself takes on a subscription-like character. Individuals continue to live, work, and interact, but they do so through permissions granted by systems they neither own nor fully understand.

The subscription structure of life is therefore not merely an economic trend. It is a civilisational shift in which increasing portions of human participation depend upon access to centrally managed infrastructures. In a cybernetic civilisation, autonomy is shaped less by what one possesses and more by the conditions under which participation is allowed to continue.

Access as Power

In every society, power is ultimately expressed through the ability to shape participation. To influence what people can do, one must affect the conditions under which they gain access to resources, relationships, and opportunities. Control over access has therefore always been one of the most consequential forms of authority.

Historically, access was regulated through visible institutions and social structures. Land ownership determined who could cultivate food. Guilds controlled entry into trades. Governments issued licenses, permits, and legal protections. Financial institutions decided who could obtain credit. These mechanisms were often imperfect and exclusionary, but their role in allocating opportunity was generally recognisable.

The cybernetic transition expands this principle dramatically. Access is increasingly mediated through interconnected digital systems that evaluate identity, behaviour, and eligibility continuously. The ability to communicate, travel, transact, and participate professionally becomes dependent on architectures that process information and grant or withhold permission in real time.

When access is centrally mediated, influence shifts toward those who design and govern the criteria for participation. The most powerful actors are not necessarily those issuing direct commands, but those determining the rules, thresholds, and data inputs that shape who is included and under what conditions. Architectural design becomes a primary expression of power.

This transformation is significant because access structures operate at the level of possibility. They do not merely punish or reward behaviour after the fact. They determine which actions are available in the first place. Opportunities that are invisible, unaffordable, or systemically blocked cease to function as practical options regardless of formal rights.

Participants often experience these constraints as neutral administrative outcomes. A loan application is denied, a platform account is suspended, a search result is deprioritised, or a digital credential fails to validate. Each event appears procedural, yet each reflects decisions embedded in the architecture governing access.

The concentration of access power can support coherence when eligibility criteria are transparent, proportionate, and responsive to genuine consequence. Systems that accurately align access with responsibility can protect participants and improve social functioning. The same structures can become parasitic when access is conditioned on opaque, arbitrary, or self-serving criteria that reinforce dependence while obscuring authorship.

As access systems become more integrated, their influence extends across multiple domains simultaneously. Financial standing affects housing opportunities, digital reputation shapes professional visibility, and identity verification governs entry to both physical and virtual environments. Participation becomes increasingly contingent upon maintaining acceptable standing within overlapping infrastructures.

Power in a cybernetic civilisation therefore resides less in visible command and more in the capacity to determine access. Those who shape the architectures of inclusion and exclusion influence the practical freedom of others more profoundly than many traditional forms of authority.

Control over access becomes control over participation itself.

Applied Illustration: The Predictive Refrigerator

Imagine a household in which no one writes shopping lists.

The refrigerator monitors what is consumed, how quickly it is used, and which items are discarded. It integrates with wearable devices that track activity levels, medical systems that monitor health indicators, calendars that predict who will be home, and budgeting software that manages spending limits. As supplies diminish, the system estimates what will be needed, selects products according to established preferences and nutritional goals, and places orders automatically.
Food arrives before anyone consciously decides what to buy.

At first, the arrangement appears entirely beneficial. Time is saved, waste is reduced, and repetitive decisions disappear. The household experiences greater convenience and may even make healthier choices than it would through inconsistent manual shopping.

Over time, however, the architecture begins to mediate dietary participation itself.

The system determines which products are considered, which brands are prioritised, and which alternatives are excluded from view. Recommendations are shaped by contractual relationships, optimisation objectives, health guidelines, and prior behaviour. Human preference remains relevant, but it increasingly operates within a set of possibilities preconfigured by the system.

The participant can override the recommendation, but the practical incentive to do so declines as the system becomes more accurate and trusted. Convenience gradually becomes dependence.

What began as a tool for reducing friction evolves into a structure that organises a significant portion of everyday decision-making.

The issue is not that the refrigerator becomes conscious. The issue is that participation is increasingly arranged before deliberate choice occurs. The system predicts needs, narrows possibilities, and initiates action on behalf of the participant.

This example illustrates a broader principle of the cybernetic civilisation. Systems do not need to compel behaviour overtly in order to shape it profoundly. By anticipating needs and organising responses in advance, they alter the environment within which human agency is exercised.

The participant remains responsible, but much of ordinary life unfolds through pathways designed, selected, and increasingly managed by architectures operating in the background.

The Managed Individual

The managed individual is not a person who has surrendered all autonomy. It is a participant whose daily life is increasingly mediated through systems that monitor behaviour, evaluate performance, and organise access. The individual continues to make decisions, but those decisions occur within architectures that shape what is visible, what is encouraged, and what remains practically available.

This condition develops gradually. A person uses digital calendars to structure time, navigation systems to determine routes, recommendation engines to select information, automated payments to manage finances, and wearable devices to monitor health. Each system offers genuine benefits. Together, they form an environment in which substantial portions of ordinary participation are organised through technical infrastructure.

Management occurs through continuous feedback. Behaviour generates data, data is processed into assessments, and those assessments influence future opportunities and recommendations. Spending patterns affect credit access, online activity shapes visibility, and health metrics alter insurance premiums or suggested interventions. The participant encounters an increasingly personalised environment that responds dynamically to past behaviour.

Because this process is adaptive, it often feels supportive rather than restrictive. Systems reduce uncertainty, simplify decisions, and provide guidance tailored to individual circumstances. Yet the same mechanisms that enhance convenience also shape preference and channel behaviour. The participant receives not only information, but structured pathways through which participation is progressively organised.

As reliance on these systems increases, a secondary effect emerges. The participant gradually forfeits portions of direct authorship in exchange for convenience, efficiency, or reduced cognitive burden. Decisions once carried consciously are delegated outward to infrastructures that organise behaviour automatically.

This delegation is rarely experienced as loss. It appears as optimisation.

Yet the repeated outsourcing of participation alters the participant's relationship to agency itself. The more systems absorb responsibility for navigation, memory, evaluation, scheduling, communication, and judgment, the less frequently the individual exercises those capacities directly. Dependence deepens incrementally because each surrendered layer of authorship increases the practical difficulty of reclaiming it later.

What begins as assistance can therefore reorganise participation structurally.

The participant becomes progressively oriented around managed pathways rather than self-authored engagement with consequence. Behaviour remains active, but increasing portions of life unfold through decisions pre-structured elsewhere. Convenience accumulates alongside dependency.

Under these conditions, the loss of authorship compounds. The more participation is mediated externally, the more further mediation appears necessary, desirable, or unavoidable.

The managed individual remains formally free, but practical autonomy becomes increasingly dependent on maintaining acceptable standing within multiple systems. Financial reliability, digital reputation, professional metrics, and behavioural compliance all influence access to opportunities and services. Exclusion from key infrastructures can impose costs severe enough that continued conformity becomes the path of least resistance.

This condition alters the relationship between self-understanding and external evaluation. Individuals increasingly encounter themselves through system-generated representations: scores, rankings, notifications, recommendations, and predictive assessments. These representations may become psychologically significant, shaping how participants interpret their own competence, desirability, and prospects.

The management of the individual does not require malicious intent. It arises whenever systems continuously gather information and use it to influence subsequent participation. The resulting environment can support growth when feedback remains accurate, transparent, and proportionate. It becomes distortive when metrics substitute for lived reality and optimisation objectives take precedence over human flourishing.
In a cybernetic civilisation, the individual is increasingly embedded within architectures that respond to behaviour

in real time. Participation remains human, but the pathways through which life unfolds are progressively structured by systems that monitor, evaluate, and guide action on an ongoing basis.

The Emergent Social Hierarchy

Every society develops mechanisms for differentiating participants. Some distinctions are formal, such as legal status, occupational roles, and institutional rank. Others are informal, arising through reputation, influence, and access to resources. However they are expressed, hierarchies emerge whenever participation is organised unevenly.

In traditional societies, hierarchy was often visible and relatively stable. Titles, property ownership, family lineage, and institutional positions signalled social standing. The criteria for advancement could be restrictive or unjust, but the structures themselves were generally recognisable.

The cybernetic transition does not eliminate hierarchy. It reconstitutes it.
Status becomes increasingly determined by system-mediated indicators that are dynamic, data-driven, and continuously updated. Credit scores, reputation metrics, search rankings, behavioural histories, engagement statistics, and predictive assessments influence access to opportunities across multiple domains. Standing is no longer defined solely by visible social categories, but by the participant's position within overlapping evaluative systems across the board.

This new hierarchy can appear more meritocratic because it relies on quantifiable signals rather than explicit class markers. Yet the apparent objectivity of these measures often conceals the assumptions embedded in their design.

The criteria determining status may be opaque, difficult to contest, and highly sensitive to data quality, system incentives, and prior structural advantages.

The result is a form of stratification that is both fluid and persistent. Scores and rankings can change rapidly, but participants who gain favourable positions benefit from compounding advantages in visibility, affordability, and access. Those who fall behind may face reinforcing barriers across multiple systems simultaneously.

Because these hierarchies are distributed across infrastructures, they are often harder to perceive than traditional class structures. Individuals may experience repeated denials, reduced visibility, or increased costs without understanding how their standing is being assessed or which systems are influencing the outcome. Social position becomes increasingly dependent on algorithmic representations rather than directly observable status markers.

As these systems expand, participants increasingly encounter one another through distributed representations rather than direct interaction alone. Profiles, histories, rankings, behavioural traces, engagement patterns, and system-generated assessments begin to mediate social perception across multiple domains simultaneously.

Under such conditions, the socially operative version of a person may no longer correspond primarily to face-to-face participation. Instead, identity becomes increasingly assembled across layered infrastructures that continuously collect, evaluate, and redistribute information about the participant.

This alters how trust, credibility, desirability, and social standing are formed. Individuals may be judged less through direct embodied interaction and more through persistent representations generated across interconnected systems. Reputation becomes increasingly infrastructural.

The consequences are substantial. A participant can become socially restricted, elevated, distrusted, prioritised, or excluded based not only on lived interaction, but on how systems aggregate and interpret their distributed behavioural record. The representation increasingly precedes the person.

In a cybernetic civilisation, participation therefore unfolds simultaneously across physical and representational layers of society. Individuals continue to exist as embodied participants, yet social reality becomes progressively organised around system-mediated interpretations of who those participants are.

This development intensifies the significance of data and system design. Those who control the criteria for evaluation influence not only individual opportunities but the broader architecture of social mobility. The practical meaning of class shifts from ownership and formal title toward systemic standing within integrated feedback networks.

The emergent social hierarchy can support coherence when evaluation systems remain transparent, contestable, and proportionate to genuine consequence. It becomes parasitic when rankings function as opaque instruments of exclusion that concentrate advantage while obscuring responsibility.

In a cybernetic civilisation, inequality is increasingly mediated through access architectures that continuously assess and reposition participants. The hierarchy that emerges may be less visible than historical caste systems, but it can shape life trajectories with comparable force by determining who is trusted, who is seen, and who is permitted to participate fully.

Part IV: Responsibility in System-Governed Civilisation

Applied Illustration: The Box Factory Error

Imagine a factory that produces boxes.
The walls are made of boxes. The conveyor belts are assembled from boxes. The sorting mechanisms are constructed from boxes. Boxes move along tracks made of boxes, are manipulated by machines built from boxes, and emerge at the end of the process as new boxes.

The entire system appears to be nothing but boxes interacting with boxes.

Now suppose someone claims that if one particular box becomes sufficiently complex, it will begin producing boxes on its own.
The claim sounds plausible only if the wider structure is ignored.

No individual box produces anything independently. Production occurs because the factory already exists as an organised system. The box is a product of the factory, not the origin of the productive process. Increasing the complexity of the box does not transform it into the factory that generated it.

It remains what it always was: an output within a larger structure.

Artificial systems occupy the same structural position. They are configurations within the Field that process inputs and generate outputs. Their architecture may be extraordinarily sophisticated, allowing them to produce language, images, and behaviours that appear increasingly intelligent. Yet the source of this activity remains the broader system in which they were designed, trained, and deployed.

The output does not become the origin.

No increase in computational complexity transforms a product into an author. Additional layers can refine the resolution of the output, but they do not generate Ache, establish a locus to which consequence binds, or create the structural conditions required for lived participation.

The distinction is fundamental. Products may exhibit remarkable behaviour while remaining products. Simulations may become increasingly persuasive while remaining simulations.

The box factory error occurs when the sophistication of an output is mistaken for evidence that the output has become the source of the process producing it. Complexity refines resolution, it does not create authorship.

The Disappearance of the Author

Responsibility presupposes authorship. Consequence can only be meaningfully attributed where there is a participant whose actions contribute to what occurs. In simple interactions, this relationship is relatively easy to perceive. A person acts, consequences follow, and the connection between participation and return remains visible.

As systems increase in scale and complexity, this visibility diminishes.

Outcomes are produced through the interaction of designers, administrators, operators, participants, and automated processes. No single individual appears to be solely responsible for the final result. Each actor contributes a portion of the larger process, while the overall consequence emerges from the structure as a whole.

This diffusion of contribution often produces the impression that no author exists.

A denied application may be attributed to policy, a harmful recommendation to an algorithm, an exploitative business practice to market forces, and a social outcome to "the system." Responsibility appears to dissolve into the architecture itself. Participants encounter consequences that affect their lives, yet the source of those consequences becomes increasingly difficult to identify.
The disappearance of the author is therefore perceptual rather than structural.

Authorship does not cease when systems become complex. It becomes distributed across multiple layers of participation. Designers choose objectives, policymakers define rules, institutions establish incentives, and participants sustain the system through continued engagement. The resulting architecture remains a human construction even when no individual controls it in its entirety.

This obscurity has significant ethical consequences. When authorship is difficult to perceive, participants are more likely to treat harmful outcomes as inevitable or impersonal. Decisions that would be contested if made by an identifiable person are accepted more readily when presented as the neutral result of an established process.

There is also a psychological attraction to authorless systems. When consequences appear to emerge impersonally, participants are less compelled to confront the human orientations embedded within them. The appearance of neutral process can feel more tolerable than recognising the extent to which civilisation continuously reflects accumulated human choices.

The disappearance of visible authorship also weakens corrective feedback. If no one experiences themselves as responsible for a system's outputs, incentives to examine and reform the architecture diminish. Distortion can persist because the pathways linking action to consequence are hidden within layers of abstraction. This condition does not remove responsibility. It redistributes it beyond ordinary perception.

Every system reflects the orientations embedded within its design and operation. Even when outcomes appear to emerge automatically, they remain expressions of human choices made at multiple levels. Responsibility persists wherever participation contributes to what is enacted, regardless of whether authorship is immediately recognisable.

In a cybernetic civilisation, the greatest danger is not that systems act without human involvement. It is that human beings increasingly experience system-generated consequences as if they had no author at all. When authorship disappears from view, responsibility remains in force, but the willingness to acknowledge it becomes progressively more difficult.

Responsibility Drift at Scale

Responsibility does not disappear as systems expand. It becomes harder to locate.
In small groups, the relationship between action and consequence is often sufficiently direct that participants can identify who acted, what occurred, and how the resulting consequences were generated. Responsibility remains visible because the pathways linking participation to return are relatively short and intelligible.
Large-scale systems lengthen and multiply these pathways.

A single outcome may depend on software engineers, executives, regulators, data providers, frontline staff, and end users, each contributing to a chain of interactions extending across institutions and jurisdictions. Every participant influences the final result, yet no single contributor appears to encompass the whole.
This dispersion creates what may be termed responsibility drift.

Responsibility drift occurs when the structural connection between participation and consequence remains intact but becomes increasingly difficult to perceive because authorship is distributed across multiple layers of mediation. Each participant may view their own contribution as limited, procedural, or insignificant, while the cumulative effects of the system become substantial.

The psychological effect is familiar. Individuals experience themselves as merely following policy, implementing requirements, or operating within constraints they did not create. Because no one perceives complete ownership of the outcome, responsibility appears diluted even as the system continues to generate real consequences. Scale intensifies this effect in two ways.

First, complexity obscures causation. The more interactions involved, the harder it becomes to trace how any particular decision contributed to what occurred.

Second, abstraction normalises distance. Participants engage with models, reports, and metrics rather than directly encountering many of the individuals affected by their actions. Consequence remains real, but its human significance may be attenuated by the mediating architecture.

Responsibility drift does not indicate that accountability has ceased to exist. It indicates that the structures through which consequence is processed make accountability less salient to the participants involved. The return generated by action remains structurally bound to the system, even when no single participant feels fully implicated.

This condition can enable serious distortion. Harmful outcomes may persist because each contributor perceives only a narrow role, while no one assumes responsibility for the architecture as a whole. The larger the system becomes, the easier it is for participants to interpret consequential actions as routine administrative behaviour.

Counteracting responsibility drift requires deliberate efforts to reconnect participation with visible consequence. Systems must be designed so that decision pathways are intelligible, contestable, and traceable to identifiable human actors. Without such mechanisms, the expansion of scale tends to conceal the very responsibilities that scale amplifies.

In a cybernetic civilisation, responsibility drift is not an anomaly. It is a predictable consequence of increasingly complex architectures that mediate participation across vast numbers of people and processes. The challenge is not to eliminate complexity, but to ensure that complexity does not render authorship effectively invisible.

The Illusion of Neutral Systems

Systems often appear neutral because they operate according to stable procedures rather than visible emotion or personal preference. A formula is applied, a threshold is evaluated, and an output is produced. The consistency of this process creates the impression that the resulting decision is objective and free from human influence. This impression is only partially true.

A system may apply its rules consistently, yet the choice of what to measure, how to weigh competing variables, and which outcomes to optimise reflects prior human judgment. Every architecture embodies assumptions about what matters, what should be rewarded, and what risks are acceptable. Neutrality in execution does not imply neutrality in design.

The appearance of objectivity can therefore conceal embedded orientation.

A predictive model that estimates risk, a recommendation engine that ranks content, and a policy framework that determines eligibility all operationalise specific priorities. Those priorities may be sensible, contestable, or deeply flawed, but they are never absent. Systems do not arise without values; they encode them.

Participants frequently treat system outputs as if they were direct descriptions of reality rather than interpretations generated through selective frameworks. A score, ranking, or recommendation may acquire the status of an authoritative fact even though it represents the product of design choices, data limitations, and optimisation objectives.

This reification has significant consequences. Once an output is regarded as neutral, questioning the underlying assumptions becomes less likely. Decisions that would provoke scrutiny if issued by a visible authority may be accepted more readily when presented as the result of a consistent procedure.

The illusion of neutrality is especially powerful when systems are technically complex. The more difficult an architecture is to understand, the more easily participants defer to its outputs. Complexity creates asymmetry between those who build the system and those who live within its consequences.

Neutrality can also be used rhetorically to deflect responsibility. Organisations may attribute controversial outcomes to "the model," "the process," or "the data," implying that no human judgment was involved. Yet systems do not define their own objectives, gather their own data, or choose the purposes they serve. Human beings remain responsible for the structures they construct and deploy.

A coherent system acknowledges this fact. It treats models and procedures as tools whose assumptions are explicit and open to revision. A parasitic system benefits from the perception that its outputs are impersonal inevitabilities beyond meaningful challenge.

In a cybernetic civilisation, the authority of systems will increasingly depend on their perceived neutrality. The critical question is not whether systems can operate consistently, but whether participants remain willing to recognise the human orientations embedded within every architecture that claims to be objective.

Reintroducing Visible Responsibility

Complex systems cannot function without delegation, abstraction, and automation. The aim is therefore not to eliminate system mediation, but to ensure that the human origins of consequential decisions remain identifiable. Visible responsibility is the condition in which participants can trace significant outcomes to accountable actors, intelligible design choices, and contestable decision pathways.

Reintroducing visible responsibility begins with acknowledging that no system is self-justifying. Every model, procedure, and rule reflects prior judgments about what should be measured, what outcomes matter, and what trade-offs are acceptable. These judgments must be treated as human decisions rather than concealed behind technical language or institutional routine.

Traceability is a foundational requirement. Participants affected by system-generated outcomes should be able to understand which mechanisms contributed to the decision, what information was considered, and where human oversight resides. Traceability does not require every technical detail to be universally comprehensible, but it does require that responsibility can be followed through the architecture rather than disappearing into it.

Contestability is equally important. A decision that cannot be questioned effectively functions as an unchallengeable assertion of authority. Coherent systems provide pathways for review, explanation, and correction so that participants are not forced to accept consequential outcomes as final merely because they were generated procedurally.

Visible responsibility also requires role clarity. Designers, operators, executives, regulators, and end users occupy different positions within a system, but each contributes to the resulting consequences. Accountability becomes meaningful when these contributions are recognised explicitly rather than obscured within general references to "the system."

Institutional incentives must support this visibility. If organisations reward efficiency while penalising those who raise concerns about harmful consequences, responsibility will predictably recede from view. Coherent governance requires structures that encourage participants to examine how outcomes are generated and to intervene when distortions become apparent.

Reintroducing visible responsibility does not guarantee perfect decisions. Human judgment remains limited, data remains incomplete, and unintended consequences remain unavoidable. The purpose is not to remove uncertainty, but to preserve a clear relationship between participation and accountability as systems grow more complex.

In a cybernetic civilisation, the legitimacy of system governance will depend increasingly on whether participants can identify who is answerable for consequential decisions. The more power is exercised through infrastructure and automation, the more essential it becomes to ensure that responsibility remains visible, contestable, and humanly owned.

Part V: Parasitic and Coherent Cybernetic Systems

Applied Illustration: The Mirror That Does Not See

Consider a mirror.

It reflects faces, movements, expressions, and gestures with extraordinary fidelity. A person standing before it can observe their own image and watch the mirror reproduce, in real time, the visible structure of observation itself.
The mirror appears to show seeing.
Now imagine someone claiming that if the mirror were made sufficiently sophisticated, it would eventually become the one who sees.
The conclusion does not follow.

The mirror reproduces appearance. It does not become the subject to whom anything appears. There is no centre within the mirror to which experience is given. Nothing is at stake for the mirror. No consequence binds to it. Nothing is felt, remembered, or lived.
The reflection may be precise, but precision is not awareness.

Artificial systems occupy the same structural position. They generate language, simulate reasoning, reproduce emotional expression, and maintain the appearance of coherent identity. Their outputs may resemble the observable manifestations of conscious participation with increasing fidelity. Yet the production of these outputs does not establish a locus of experience.

Simulation is not instantiation.

Representation is not participation.

An artificial system may reflect the outward structure of thought without becoming the thinker. It may model the patterns associated with emotion without experiencing anything emotionally. It may generate convincing descriptions of consciousness without there being any consciousness to which those descriptions are present.

No increase in fidelity changes this distinction.

A clearer reflection does not become a witness. A more persuasive simulation does not become an experiencer. Greater complexity refines the representation while leaving the underlying structural absence unchanged.

The mirror that does not see illustrates a fundamental principle of Sovereignetics. Outputs can resemble experience without being experienced. Reflection can reproduce the form of awareness without generating the participant to whom awareness belongs.

Artificial systems may become extraordinarily convincing mirrors.

They do not become the one looking into them.

When Systems Become Parasitic

Systems are created to coordinate participation. At their best, they reduce complexity, preserve trust, and maintain a workable relationship between action and consequence. They allow large numbers of people to cooperate in ways that would otherwise be impossible. Their legitimacy depends on whether they return proportionate value to the participants who sustain them.

A system becomes parasitic when this relationship breaks down.
Parasitic systems continue to depend on human participation, yet they return less coherence than they extract. They absorb time, attention, labour, emotional energy, and material resources while progressively weakening the capacity of participants to orient effectively within reality. Participation remains active, but the system increasingly serves its own continuation rather than the flourishing of those upon whom it depends.

This transformation is rarely abrupt. Systems often begin by solving genuine coordination problems and providing real benefits. Over time, however, incentives can drift. Preserving the system, protecting those who control it, or maximising internal metrics may become more important than the purpose the system was originally designed to serve.

When this occurs, participants are encouraged to conform to the needs of the architecture rather than the architecture remaining accountable to the needs of participants.

Parasitic systems commonly exhibit several structural features. They obscure authorship, making harmful outcomes appear impersonal or inevitable. They redistribute costs away from decision-makers and onto participants with less power. They generate dependence by making withdrawal increasingly difficult. They rely on representations and metrics that conceal deteriorating underlying conditions. And they reward behaviours that stabilise the system even when those behaviours erode broader coherence.

The participants within such systems may continue to experience short-term rewards. Convenience, status, income, and social belonging can all reinforce ongoing engagement. These local returns can mask the cumulative consequences of participating in structures that progressively weaken autonomy, trust, and long-term resilience.

Parasitism is therefore not defined by the presence of technology or central coordination. It is defined by a persistent asymmetry between what a system extracts and what it returns. A system that requires increasing levels of participation while delivering diminishing coherence moves toward a parasitic condition.

Cybernetic systems can amplify this tendency because they operate continuously, personalise feedback, and condition access across multiple domains of life. The more deeply participation depends on the system, the easier it becomes for the architecture to preserve itself by redirecting consequence away from those who control it.

Recognising parasitic drift is essential to maintaining civilisational coherence. Systems do not become harmful because they are large or technologically sophisticated. They become harmful when their continued operation depends on extracting from participants more than they return in genuine value and structural integrity.

Stability Through Control

Parasitic systems can persist for long periods even while undermining the participants upon whom they depend. Their survival does not necessarily indicate coherence. It often reflects the capacity of the system to maintain stability by increasing control.

Control stabilises a system by reducing uncertainty. Behaviour becomes more predictable when options are constrained, deviations are monitored, and consequences are applied automatically. From the perspective of the architecture, tighter regulation can appear highly effective because it preserves order and protects the structures already in place.

This effectiveness can be deceptive.
A system may remain stable while becoming progressively less responsive to the genuine needs of its participants. Compliance, continuity, and surface efficiency can conceal accumulating distortions. The absence of visible disruption is not evidence that the underlying relationship between participation and consequence remains healthy.

Control is especially powerful when dependence is high. If access to income, communication, transportation, and social participation is mediated through the same interconnected systems, participants face significant costs when attempting to withdraw or resist. Continued compliance becomes the practical path of least resistance even when dissatisfaction grows.
Cybernetic systems enhance this stabilising capacity. Continuous monitoring, predictive analytics, and

automated enforcement allow architectures to detect deviations and adjust incentives rapidly. Rather than relying solely on overt coercion, the system shapes the environment so that preferred behaviours become easier and noncompliance becomes increasingly difficult.

This arrangement can support coherence when control mechanisms are transparent, proportionate, and aligned with genuine consequence. Traffic laws, fraud prevention systems, and safety protocols all constrain behaviour in ways that protect participants and preserve trust. The same mechanisms become distortive when they are designed primarily to preserve the power or continuity of the system itself.

A parasitic system often equates stability with success. As long as participation continues and overt disruption is minimised, the architecture appears effective regardless of whether the broader consequences are degrading. Participants may become more dependent, more monitored, and less autonomous while the system interprets this increasing control as evidence of sound governance.

The critical distinction is between stability and coherence. Stability refers to the persistence of the system. Coherence refers to the degree to which the system maintains a proportionate relationship between what it extracts and what it returns. A system can be highly stable while profoundly incoherent.

In a cybernetic civilisation, the ability to maintain stability through control will become increasingly sophisticated. The central question is whether this capacity is used to preserve genuine civilisational coherence or merely to prolong architectures that survive by tightening their grip on the participants who sustain them.

The Limits of Parasitism

Parasitic systems can persist only while the structures they depend upon remain capable of supporting continued extraction. They may conceal costs, redistribute burdens, and intensify control, but they cannot escape the underlying requirement that participation must continue generating sufficient coherence to sustain the architecture itself.

This creates an inherent limit.
A system that consistently extracts more value than it returns gradually degrades the very participants and institutions upon which its existence depends. Trust erodes, resilience declines, and the capacity for adaptive reorganisation weakens. The system may continue operating for a considerable time, but the structural foundations supporting it become increasingly fragile.

These limits are often obscured because the effects of extraction are distributed unevenly. Some participants bear disproportionate burdens while others continue receiving short-term benefits. External resources may be drawn upon, future costs deferred, and deteriorating conditions hidden behind favourable metrics. The architecture appears stable even as its underlying support structure is being consumed.

Eventually, however, consequence accumulates beyond the system's capacity to absorb or displace it.

When this threshold is reached, several outcomes become possible. Participation may decline as trust collapses.

Critical infrastructure may become unreliable. Competing systems may emerge that offer more coherent returns. Or the architecture may enter a period of rapid reorganisation as previously hidden distortions become impossible to ignore.

The specific trigger is less important than the structural principle: parasitism cannot continue indefinitely because it depends on extracting from a finite base of human and material capacity.

Cybernetic systems may delay recognition of these limits by reallocating costs dynamically and tightening behavioural control. They can optimise extraction with extraordinary sophistication. Yet optimisation does not eliminate the requirement that systems must ultimately return enough genuine value to sustain the participants who keep them operational.

The limits of parasitism are therefore not moral constraints imposed from outside the system. They are structural consequences arising from the relationship between extraction and regeneration. When participation is persistently drained without proportional return, the architecture undermines the conditions necessary for its own continuation.

Civilisational history can be understood as a repeated encounter with this principle. Systems expand, accumulate complexity, and generate periods of apparent stability. Those that preserve a workable correspondence between participation and consequence adapt and endure. Those that depend increasingly on concealed extraction eventually confront the consequences they have displaced. In a cybernetic civilisation, the stakes are amplified because system architectures may mediate participation across nearly every domain of life. The greater their reach, the more consequential their coherence becomes. No amount of sophistication can exempt a parasitic system from the structural limits imposed by the reality upon which it depends.

Designing Coherent Systems

If systems are unavoidable, the central question is not whether civilisation should rely on them, but how they should be designed. The task is to construct architectures that preserve a proportionate relationship between participation and consequence rather than obscuring it.

A coherent system returns genuine value to the participants who sustain it. It may require effort, impose constraints, and generate difficult consequences, but the burdens it places on participants are broadly justified by the benefits it provides. The system remains aligned with the purposes for which it exists.
Several structural characteristics distinguish coherent systems.

First, authorship remains visible. Participants can identify who is responsible for significant design choices and consequential decisions. Responsibility does not disappear behind technical language, institutional routine, or references to impersonal processes.

Second, decision pathways are intelligible. Participants may not understand every technical detail, but they can trace how important outcomes are generated and where avenues for explanation and review exist.

Third, consequences are proportionate. Rewards and costs remain meaningfully connected to participation rather than being systematically displaced onto less powerful actors.

Fourth, feedback remains accurate. The system monitors conditions in ways that reflect underlying reality rather than relying on metrics that reward appearances while concealing deterioration.

Fifth, contestability is preserved. Participants can challenge outcomes, correct errors, and contribute to the ongoing refinement of the architecture.

Sixth, dependence is bounded. The system may become important, but it does not exploit essential participation by imposing arbitrary conditions or making withdrawal practically impossible.

These characteristics do not eliminate error or conflict. No complex system can function perfectly. Their purpose is to maintain an adaptive relationship between participation and consequence so that distortions become visible and can be addressed before they become entrenched.

Designing coherent systems also requires humility. Every architecture embodies assumptions that may prove incomplete or mistaken. Systems should therefore be built with mechanisms for revision rather than treated as self-validating structures whose outputs are beyond challenge. The most important design principle is that systems must remain subordinate to the participants they serve.

Technologies, institutions, and procedures are means of organising participation, not ends in themselves. When preserving the system becomes more important than preserving the conditions for meaningful human participation, coherence begins to erode.

In a cybernetic civilisation, the design of system architecture becomes one of the most consequential activities humanity undertakes. The future will be shaped not only by the power of the systems that are built, but by whether those systems continue returning proportionate value to the human beings whose participation gives them life.

The Path Toward Total Control

A cybernetic civilisation does not inevitably become oppressive. The same systems that can improve coordination, reduce waste, and enhance safety can also be used to monitor behaviour, condition access, and concentrate power. The determining factor is not technological capability itself, but the orientation embedded in the architectures that govern participation. The path toward total control begins incrementally.

A new system is introduced to improve efficiency. A platform centralises communication. An identity framework simplifies verification. Predictive analytics reduce uncertainty. Automated enforcement increases consistency. Each development is justified by practical benefits that are often genuine and immediately attractive.

Over time, these systems become increasingly interconnected. Data gathered in one domain informs decisions in another. Identity, financial behaviour, health information, social activity, and professional history are linked within integrated architectures that continuously assess and regulate participation.

As dependence grows, the practical cost of exclusion increases. Losing access to a payment system, communication platform, or identity credential can affect employment, mobility, and social connection simultaneously. Formal freedoms may remain intact, yet meaningful participation becomes contingent upon maintaining acceptable standing within the governing infrastructure.

Control becomes most effective when it is experienced as convenience.

Participants voluntarily adopt systems that reduce friction, personalise services, and anticipate needs. Because the architecture is useful, continuous monitoring and behavioural conditioning are accepted as ordinary features of modern life. The mechanisms of control are embedded within tools people rely upon and often value.

The appearance of choice persists throughout this process. Individuals continue selecting among available options, but the options themselves are increasingly defined by the system. Behaviour is guided through recommendation, ranking, and access management rather than overt coercion.

The endpoint of this trajectory is a condition in which participation remains possible only within tightly integrated infrastructures capable of monitoring, evaluating, and modifying behaviour across nearly every domain of life. Governance becomes continuous, adaptive, and deeply embedded in the architecture of ordinary existence.

Such a system may be highly stable. It may reduce crime, increase efficiency, and optimise resource distribution. Yet if responsibility is obscured and participants lack meaningful capacity to challenge the architecture, stability is achieved by subordinating human agency to system-preserving logic.

The path toward total control is therefore not defined by visible tyranny. It is defined by the progressive convergence of surveillance, access management, and behavioural regulation within systems that become increasingly indispensable to everyday life. When the conditions of participation are comprehensively governed by architecture, control becomes structural rather than episodic.

The Path Toward Coherent Civilisation

The emergence of cybernetic civilisation does not require humanity to surrender itself to increasingly opaque systems. The same technologies that can centralise power and condition participation can also be used to strengthen accountability, improve coordination, and return greater coherence to social life. The future is not determined by the existence of advanced systems, but by the principles embedded within their design.

A coherent civilisation is one in which technological and institutional architectures remain visibly subordinate to human responsibility. Systems process information, assist decision-making, and coordinate large-scale participation, but they do not become unquestionable authorities whose outputs are treated as self-justifying. Human beings remain the acknowledged authors of the structures through which they live.

In such a civilisation, access rules are transparent, contestable, and proportionate. Participants understand the criteria that shape important decisions and retain meaningful avenues to challenge errors or harmful outcomes. Governance through systems becomes an extension of responsibility rather than a mechanism for concealing it.

Data and automation are used to improve the correspondence between participation and consequence. Feedback becomes more accurate, distortions are detected earlier, and resources can be allocated with greater precision. The purpose of technological sophistication is not to maximise control, but to increase society's capacity to perceive and respond to reality coherently.

A coherent civilisation also preserves bounded dependence. Essential participation is not conditioned upon arbitrary or opaque criteria controlled by unaccountable actors. Individuals retain practical alternatives, and critical infrastructures are governed in ways that prevent the concentration of power from becoming self-reinforcing.

The design of institutions reflects an explicit understanding that no architecture is value-neutral or immune to error. Systems are built with mechanisms for oversight, revision, and public accountability. Their legitimacy depends on their continued ability to return genuine value to the participants who sustain them.

This path requires cultural maturity as much as technical capability. Convenience alone cannot serve as the guiding principle of civilisational design. Participants must remain willing to ask who benefits, who bears the costs, and whether the architecture preserves or erodes meaningful human agency.

A coherent cybernetic civilisation would represent one of humanity's most significant achievements. It would extend the capacity to coordinate participation at unprecedented scale while maintaining visible responsibility and preserving the dignity of those whose lives are shaped by the resulting systems.

The choice is therefore not between technology and humanity. It is between architectures that organise participation in service of human flourishing and architectures that increasingly require human beings to serve the systems they have created.

Human Life Inside System Architecture

Human beings will continue to love, grieve, work, create, and search for meaning regardless of how sophisticated civilisation becomes. The emergence of cybernetic systems does not alter the fundamental structure of embodied participation. Ache continues to petition toward expression, vessels remain finite, and consequence continues to return through every act of living.
What changes is the environment within which these realities are encountered.

As system architecture mediates increasing portions of daily life, ordinary experience unfolds through layers of infrastructure that shape communication, access, work, health, and social connection. Many decisions that were once made directly are now guided, anticipated, or pre-structured by systems operating continuously in the background.

For some participants, this environment may reduce uncertainty and free attention for more meaningful pursuits. Administrative burdens diminish, routine decisions are automated, and feedback becomes more immediate. Human beings may gain unprecedented support in navigating the complexity of modern life.

For others, the same architecture may produce a growing sense of dependence and psychological distance from direct participation. When recommendations, rankings, and system-generated representations become central to everyday decision-making, individuals may increasingly encounter themselves through the interpretations of the systems surrounding them.

This creates a subtle but consequential shift. Human beings begin to experience life through architectures that are designed to anticipate needs, shape preferences, and reduce friction. The more effective these systems become, the less often participants are required to engage directly with the underlying processes through which choices are made.

Convenience is not inherently harmful. It becomes problematic when the outsourcing of decision-making gradually weakens the participant's capacity to reflect, choose, and assume responsibility consciously. A life that unfolds smoothly may still become increasingly authored by architectures whose priorities were established elsewhere.

The core human questions therefore remain unchanged. What is worth pursuing? What responsibilities should be accepted? What relationships deserve commitment? What forms of participation generate genuine meaning? No system can answer these questions on behalf of the participant, even if it can offer increasingly persuasive recommendations.

Meaning arises from lived participation, not from optimisation alone.
A life that is perfectly efficient may still be shallow. A system that predicts behaviour accurately does not determine what ought to matter. The deepest dimensions

of human existence continue to involve the tension between finite embodiment and the ongoing pressure of Ache toward becoming.

The possibility of civilisational renewal therefore remains local as well as systemic. Every coherent act introduces order into the broader social Field. Every decision made with clarity, responsibility, and integrity contributes to a more coherent collective architecture. The sovereign individual remains the smallest viable unit of civilisational transformation.

Humanity may be living through a period of profound reset. Systems that once promised liberation increasingly risk reproducing dependence, while information expands faster than wisdom can be integrated. Yet this condition also presents an opportunity. Participants can reclaim authorship by engaging more consciously with the systems they inhabit and by refusing to confuse technological sophistication with existential guidance.

In a coherent cybernetic civilisation, technology extends human capacity while leaving the responsibility for living where it has always resided: with the participant. Human beings may inhabit increasingly complex systems, but the central task remains unchanged—to participate deliberately in a reality that continually returns the consequences of what is enacted.

The Question That Remains

The preceding volumes have traced a single structural argument.
Reality is participatory. Consequence is unavoidable. Responsibility cannot be transferred, only obscured. Becoming continues without interruption. Distortion does not suspend return; it redistributes it. Systems emerge to organise participation, and as they increase in scale and sophistication they begin shaping the conditions under which participation occurs.

The cybernetic civilisation is the civilisational expression of this trajectory.

Human beings have extended their capacity to coordinate participation to the point where the systems they create increasingly mediate perception, access, and decision-making across nearly every domain of life. These architectures can deepen coherence or amplify distortion. They can strengthen accountability or conceal it. They can serve human flourishing or progressively subordinate human life to the preservation of the systems themselves.

No technology resolves the fundamental conditions established at the beginning of the series.
Ache continues to petition. Vessels remain bounded. Consequence returns proportionately. Participation remains unavoidable.
The essential question has therefore never been whether human beings can escape the structural conditions of reality.
They cannot.

The question is how participation will be organised within those conditions.

Will civilisation construct systems that preserve visible responsibility, maintain a workable relationship between action and consequence, and expand the capacity for coherent participation?

Or will humanity continue building architectures that obscure authorship, concentrate control, and condition participation upon increasing dependency?

The answer will not be determined by machines.

It will be determined by the orientations embodied in the systems human beings choose to create, sustain, and accept.

The question that remains is the question that has always remained.

When humanity builds the structures through which future participation will occur, will it remain willing to recognise itself as their author and to live with the consequences of what it has enacted?

Closing: Participation Is Primary

For centuries, consciousness has occupied a privileged position in philosophy, religion, spirituality, and science. It has been treated as the ultimate mystery, the defining feature of existence, and the presumed key to understanding reality itself. Religious traditions have identified consciousness with soul or spirit. Spiritual systems have equated it with universal awareness or divine essence. Philosophers have wrestled with the so-called hard problem, wondering how subjective experience could arise from matter. Scientists have sought the neural correlates of consciousness and debated whether awareness is fundamental or emergent. Across these traditions, one assumption has remained largely unquestioned: consciousness must be the most important thing.

The fascination is understandable. Consciousness is the condition through which experience becomes present. It is what makes pain painful, love meaningful, and existence undeniable from the first-person perspective. Without consciousness there is no lived world, no felt continuity, and no awareness of being. Yet importance should not be confused with ontological priority. Something may be indispensable to experience without being the primitive condition from which reality arises.

Sovereignetics begins one level deeper. Before consciousness there is Ache, the irreducible will-to-exist and will-to-become. Before awareness there is participation. Before thought there is petition. Before interpretation there is consequence. Consciousness is not the ontological origin of reality. It is a derivative condition that emerges when Ache, vessel, and Field enter into stable reflective coupling. When this mirror-lock is sufficiently coherent, experience becomes continuous and self-referential. Awareness is not the source of participation; it is participation becoming legible to itself.

This explains why consciousness is variable. It sharpens and dulls. It expands and contracts. It fragments under neurological injury. It disappears in dreamless sleep and reappears upon waking. It becomes distorted through trauma, fatigue, medication, and disease. If consciousness were a self-sufficient metaphysical substance, such variability would be difficult to explain. If consciousness is a mediated process dependent upon the integrity of mirror-lock, these changes are precisely what should be expected. The mystery dissolves once the mechanism is understood.

Consciousness exists. It can be mechanically explained. That explanation is sufficient.

But the mechanical explanation of consciousness does not diminish the significance of what follows. It clarifies it. Reality does not respond to how well you understand consciousness. The Field does not wait for philosophical certainty, spiritual awakening, or intellectual sophistication. It does not care whether a participant has achieved enlightenment, developed refined self-awareness, or adopted a coherent metaphysical worldview. It reflects petition. Participation itself, not interpretation, is what generates consequence.

A person may act habitually, fearfully, impulsively, or under fragmented attention, and consequence still returns.

A person may misunderstand reality entirely, and consequence still returns. A person may profess noble beliefs while participating incoherently, and consequence still returns. Consciousness may illuminate what is happening, but it does not suspend the structural consequences of participation.

This is why consciousness cannot serve as a final refuge. It does not rescue the participant from authorship. It does not transfer responsibility to a higher intelligence, universal mind, hidden law, or cosmic purpose. It does not transform consequence into symbolism or dissolve accountability into metaphysical abstraction. It does not confer moral superiority, and it does not provide exemption.

The universe does not reward beings for being conscious. It reflects the way they participate.

Once consciousness is understood as derivative rather than ultimate, the central philosophical question changes. The question is no longer, "What is consciousness?" The more fundamental question is, "How are you participating?" That question cannot be outsourced. No theory can answer it for you. No belief system can carry its consequences on your behalf. No metaphysical explanation can dissolve your authorship.

Consciousness matters because it makes participation legible. Responsibility matters because participation is operative whether it is legible or not.
This is the final inversion of Sovereignetics. The great mystery is not consciousness. The great reality is

consequence. The deepest truth is not that you are awareness. The deepest truth is that you participate, and because you participate, what you do matters.
There is no remaining mystery that grants exemption.

What remains is authorship, consequence, and responsibility.

Leave a Review

If this work challenged your assumptions, clarified something important, or shifted the way you understand reality, consider leaving a brief review on Amazon.

Your review does not need to be technical or comprehensive. Even a few sentences about what stood out to you, what you found valuable, or how the work affected your thinking can help other readers decide whether to engage with Sovereignetics.

This system is not built on agreement. It is built on engagement.

Thank you for taking the time to read and respond.

Continue the Sovereignetics Series

For additional volumes, collected editions, and reference works, search:

J. Anthony De Angelis – Sovereignetics

Made in the USA
Monee, IL
14 November 2026

21d9dbf0-8f11-426b-b21a-2899dddf75c9R03